Y0-ABZ-075

LESS THAN NOTHING
IS REALLY SOMETHING

LESS THAN NOTHING
IS REALLY SOMETHING

A YOUNG MATH BOOK

BY ROBERT FROMAN

ILLUSTRATED BY DON MADDEN

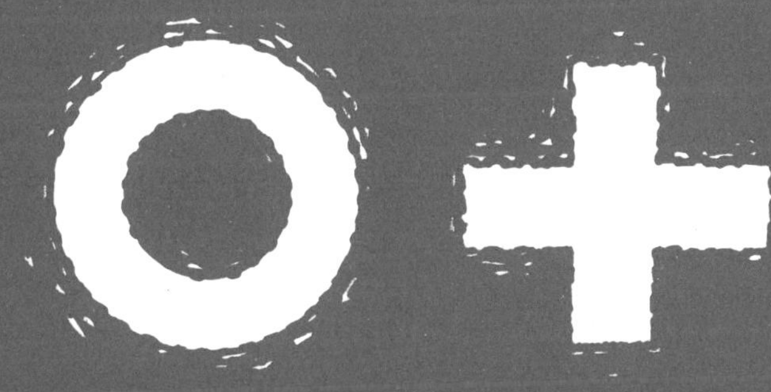

THOMAS Y. CROWELL COMPANY NEW YORK

YOUNG MATH BOOKS

Edited by Dr. Max Beberman, Director of the Committee on
School Mathematics Projects, University of Illinois

BIGGER AND SMALLER
by Robert Froman

CIRCLES
by Mindel and Harry Sitomer

COMPUTERS
by Jane Jonas Srivastava

THE ELLIPSE
by Mannis Charosh

ESTIMATION
by Charles F. Linn

FRACTIONS ARE PARTS OF THINGS
by J. Richard Dennis

GRAPH GAMES
by Frédérique and Papy

LINES, SEGMENTS, POLYGONS
by Mindel and Harry Sitomer

LONG, SHORT, HIGH, LOW, THIN, WIDE
by James T. Fey

MATHEMATICAL GAMES FOR ONE OR TWO
by Mannis Charosh

ODDS AND EVENS
by Thomas C. O'Brien

PROBABILITY
by Charles F. Linn

RIGHT ANGLES: PAPER-FOLDING GEOMETRY
by Jo Phillips

RUBBER BANDS, BASEBALLS AND DOUGHNUTS:
A BOOK ABOUT TOPOLOGY
by Robert Froman

STRAIGHT LINES, PARALLEL LINES,
PERPENDICULAR LINES
by Mannis Charosh

WEIGHING & BALANCING
by Jane Jonas Srivastava

WHAT IS SYMMETRY?
by Mindel and Harry Sitomer

Edited by Dorothy Bloomfield, Mathematics Specialist,
Bank Street College of Education

LESS THAN NOTHING IS REALLY SOMETHING *by Robert Froman*

STATISTICS *by Jane Jonas Srivastava*

VENN DIAGRAMS *by Robert Froman*

 Published simultaneously in Canada by Fitzhenry & Whiteside Limited, Toronto.
Manufactured in the United States of America
ISBN 0-690-48862-9
0-690-48863-7 (LB)

1 2 3 4 5 6 7 8 9 10

Library of Congress Cataloging in Publication Data
Froman, Robert. *Less than nothing is really something.* (Young math books)
SUMMARY: Explains in simple terms the concept of positive and negative numbers.
1. Arithmetic—Juvenile literature. 2. Numbers, Natural—Juvenile literature. 3. Numbers, Negative—Juvenile literature. [1. Arithmetic] I. Title. QA107.F76 513'.2 72-7546 ISBN 0-690-48862-9 ISBN 0-690-48863-7 (LB)

LESS THAN NOTHING
IS REALLY SOMETHING
YOUNG MATH BOOKS

Suppose you have a penny. Then suppose you find another penny. It's easy to count the pennies you have. One, two or 1, 2.

Suppose you lose both these pennies. It's still easy to know how many pennies you have. The number is none or 0.

But suppose you have one penny, and you want to buy a piece of candy that costs two pennies. You have a friend named Laura, and she lends you a penny. You promise to pay it back to her the next day.

You run to the store, buy the piece of candy, and eat it. Yum.

Do you know how many pennies you now have?

It would not be right to say that you have 0 pennies. What about the penny that you promised to pay back to Laura? The number of pennies you now really have is less than 0.

When you subtract one from two, you can write down what you do as two minus one, or 2 – 1. When you subtract one from none, you can write down what you do as none minus one, or 0 – 1.

None minus one, or 0 – 1, is the same as plain minus one or –1. So let us call the penny you owe Laura minus one. Or you can write it down as –1.

0-1 = -1

A number that is less than 0, like -1, is called a NEGATIVE number. Negative numbers are very useful in many different ways.

They are useful in telling how cold the weather is in winter when it gets very, very cold.

Do you have an outdoor thermometer at your house? It probably looks something like this:

In summer it may show 70 degrees, which is usually written 70°. When the thermometer shows 70°, the weather is nice and warm.

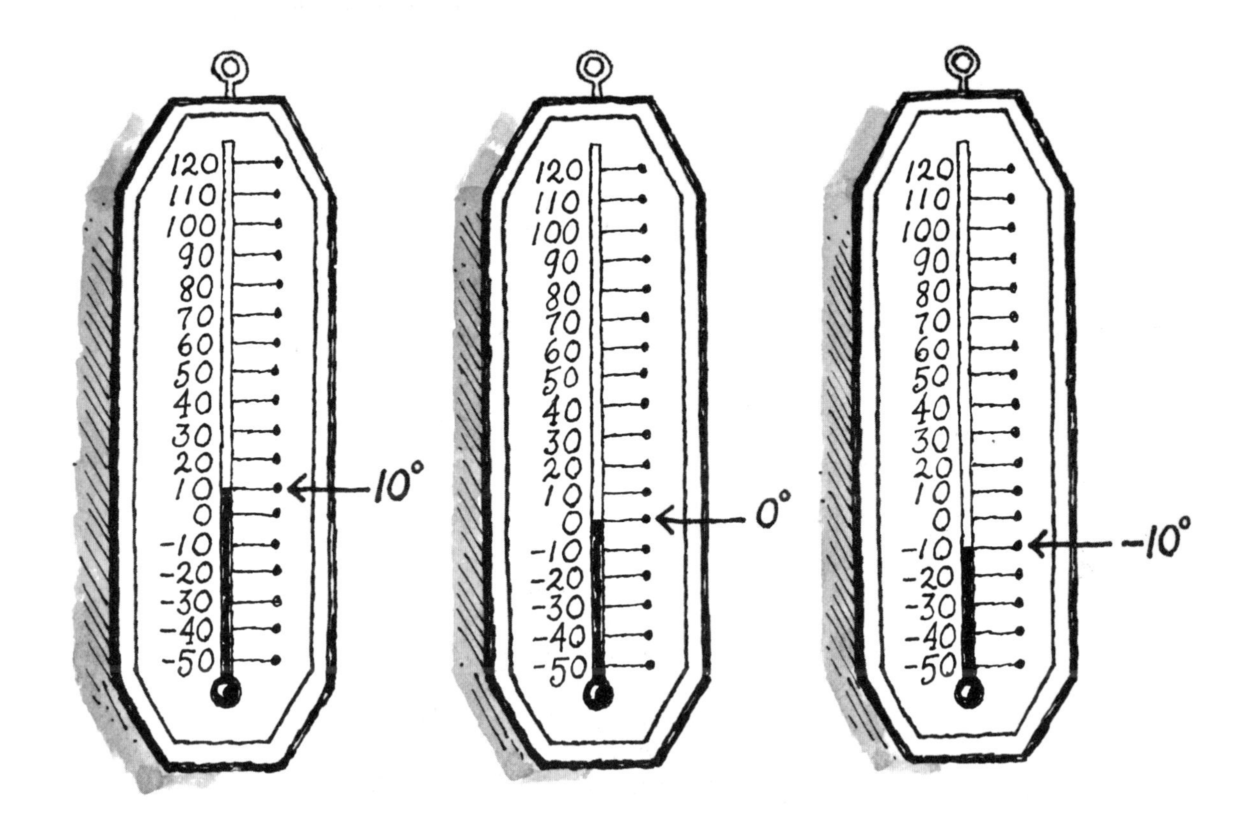

In the cold parts of the world in winter, a thermometer may show a temperature like 10°. That is quite cold. Or it may show 0°, which is even colder.

But 0° is not the coldest possible weather. In some places the temperature often goes below 0°. When it does this, you write it as a negative number. Ten degrees below zero is −10°.

125
100
75
50
25
0
-25
-50
-75
-100
-125

There are many places where the temperature goes to −30° in winter. In some places it goes as low as −70° almost every winter. Near the South Pole in Antarctica, an outdoor thermometer once showed a temperature of −125°. That is very, very cold.

Can you think of other ways to use negative numbers?

You can use them to keep track of steps down. Suppose you are in a house with a basement, and suppose there are fifteen steps leading down into the basement. You could call the first step down −1. And you could call the second step down −2. And when you are at the bottom, you could say you are at −15.

9
-10
-11
-12
-13
-14
-15

This is also the way to keep track of the depth of water in a swimming pool or a lake or an ocean. When you dive ten feet down, you are at −10 feet. When you dive another ten feet down from there, you are at −20 feet.

In the very deepest place in the ocean, the bottom is lower than −35,000 feet.

One other way to use negative numbers is to keep track of time. They are used this way when a rocket ship is launched on a trip into space.

The blastoff, which is the instant when the rockets are fired, is called 0. The seconds before blastoff are counted as $0 - 1$, $0 - 2$, and so on. The seconds after blastoff are counted as 0 plus one, 0 plus two, and so on. They usually are written $0 + 1$, $0 + 2$, and so on.

Here is the way the count goes from five seconds before blastoff to five seconds after blastoff:

The numbers to the right of 0 in this count—the ones with the plus sign—are called POSITIVE numbers.

← Negative Numbers 0 Positive Numbers →

The positive numbers and the negative numbers together with 0, which is neither positive nor negative, are parts of the NUMBER LINE. You can think of the numbers from -5 to $+5$ as names of points on the number line. Like this:

When you move to the right along this line, each number you come to is one more than the number before it.

This is easy to see when you move to the right from one positive number to the next. If you start at +1, the next number is +2, then +3, and so on.

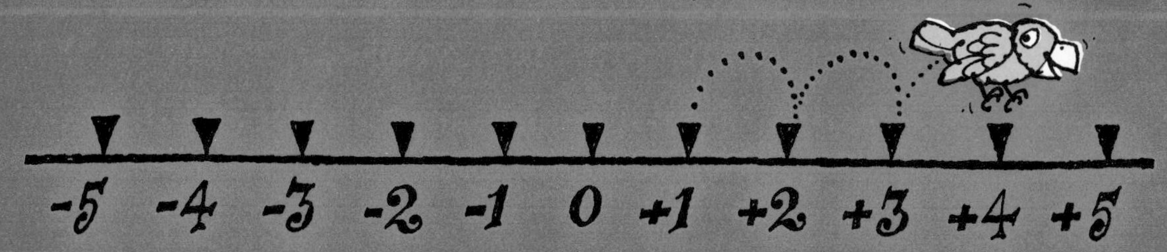

It is not so easy to see when you move to the right from one negative number to the next. If you start at −3, the next number to the right is −2. Then comes −1. Then comes 0.

Is −2 one more than −3? Is −1 one more than −2? Is 0 one more than −1?

In each case the answer is yes.

You can think of −3 as meaning that you have a debt of 3 pennies. Then −2 would mean a debt of 2 pennies, −1 a debt of 1 penny, and 0 means no debt. Each time you move a step to the right on the negative side of 0, it is as if you have paid off one penny of your debt. When you have done that, you are one penny better off.

So it is true that, wherever you start, when you move to the right along the number line, each number you come to is one more than the number before it.

Can you see what happens when you move to the left along the number line?

There is a new game you can play with negative numbers and positive numbers. It is called P.A.M. While you are playing the game, see if you can guess what P.A.M. stands for.

You don't have to buy anything to play P.A.M. Just find a smooth tabletop and a big piece of paper and a pencil to mark it with. You will also need four counters—either buttons or coins like nickels or dimes. To start with, mark the paper like this and fasten it down with tape near one end of the tabletop.

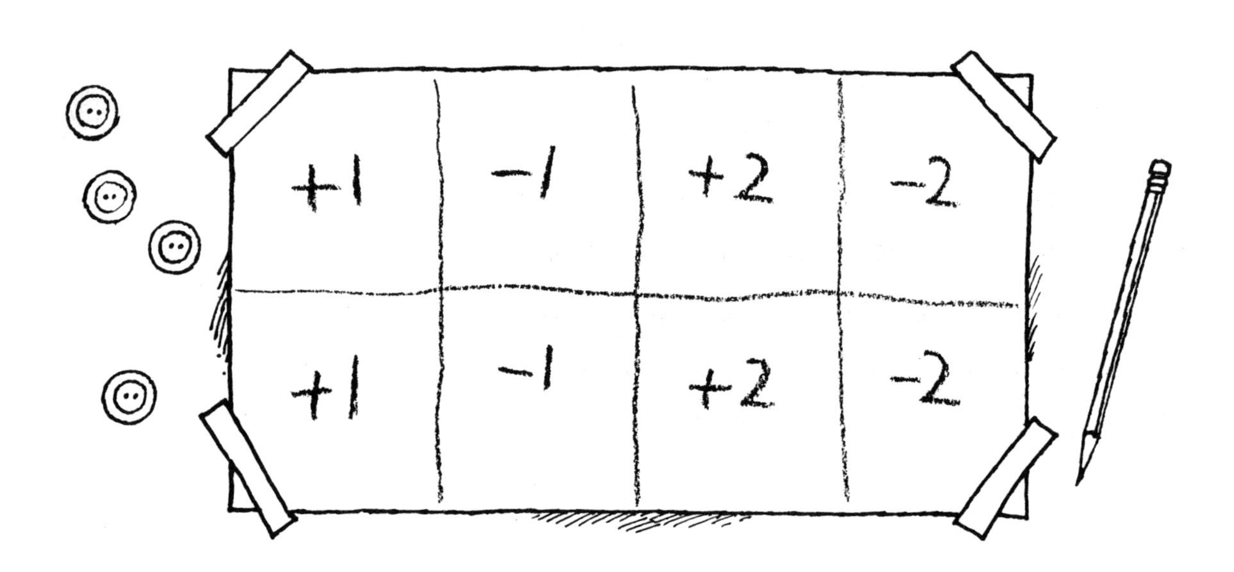

Two or more people can play. Or if there is no one around to play with, you can match your left hand against your right hand. The players stand at the end of the table farthest from the piece of paper. Each player lines up two counters on the table in front of him. Then the players take turns, flicking one counter at a time at the paper, flicking them with a fingernail. Make little marks on each counter so that you can be sure who it belongs to. And be sure that each square is big enough to hold several of the counters.

Each player's score depends on where his counters stop or where they are knocked to by other counters. The number on the square tells you how much to score for each counter in that particular square. Counters that end up off the paper score nothing. If a counter stops on a line, its score depends on where the largest part of it is.

Suppose you and your friend Laura are playing the game. Laura's counters are marked L and yours are marked Y. The counters end up like this:

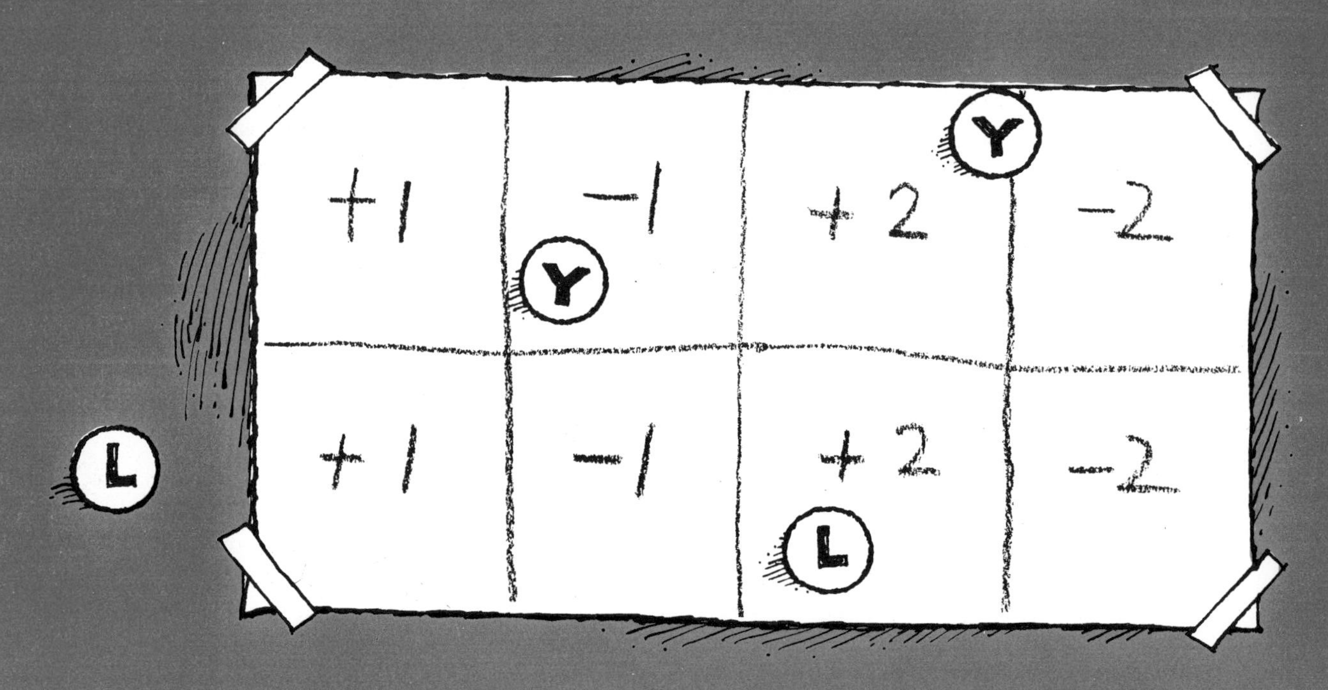

Can you tell who wins?

Laura's score is +2. Yours is +2 − 1. So it's easy to see that Laura wins the first game.

Suppose you play again, and this time the counters stop like this:

Can you tell who wins?

Laura's score is +2 − 1. Yours is +2 + 1. So it's easy to see that you win the second game.

But suppose you play again and the counters stop like this:

This time your score is $+2 - 2$. How much is that?

Laura's score is $+1 - 2$. How much is that?

This time it is not so easy to tell which score is higher.

When you have 2 and then take away 2, you are left with 0. But Laura's score is less than 0. It is as if she owes somebody one point. So your score of 0 is higher than her score of -1.

But now you play again and the counters stop like this:

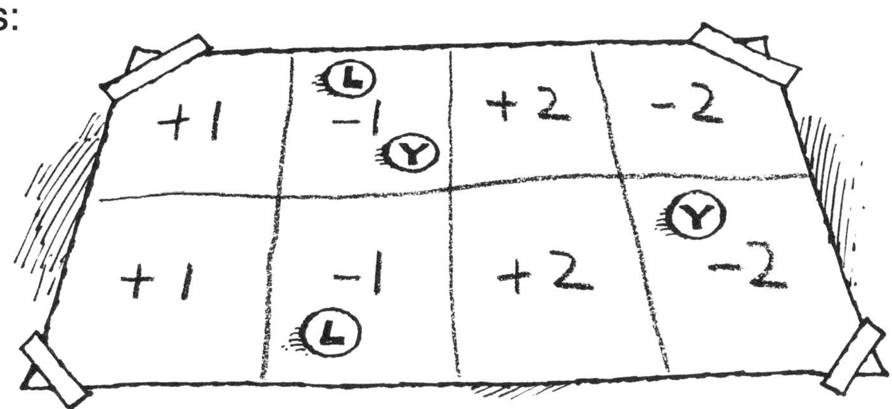

Laura's score is two times −1. Or you could write it as (−1) + (−1). Can you tell how much that is?

Your score is (−1) + (−2). Can you tell how much that is?

To add a negative number to a negative is like adding to your debt.

Laura's score is −2 and your score is −3. You could say that Laura is in debt for 2 points and you are in debt for 3 points. Laura is not as much in debt as you are, so her score is higher. She wins this time.

Your score can change in a game. If your first counter lands in a square, you can knock it out of that square with your second counter. Or maybe one of the other player's counters will knock it out.

Suppose your first counter lands in a −1 square. Your score is −1. Then along comes your second counter and knocks the first one off the paper. The second one goes off the paper, too.

Do you see what you have done here?

You have subtracted -1 from your score. It is as if you owed somebody a debt of one point, and you paid the debt. That wipes it out. Your score changes from -1 to 0.

But suppose one of your counters is in a -1 square and the other is in a -2 square. This makes your score $(-1) + (-2)$, which amounts to -3. And suppose one of Laura's counters then knocks your counter in the -1 square off the paper.

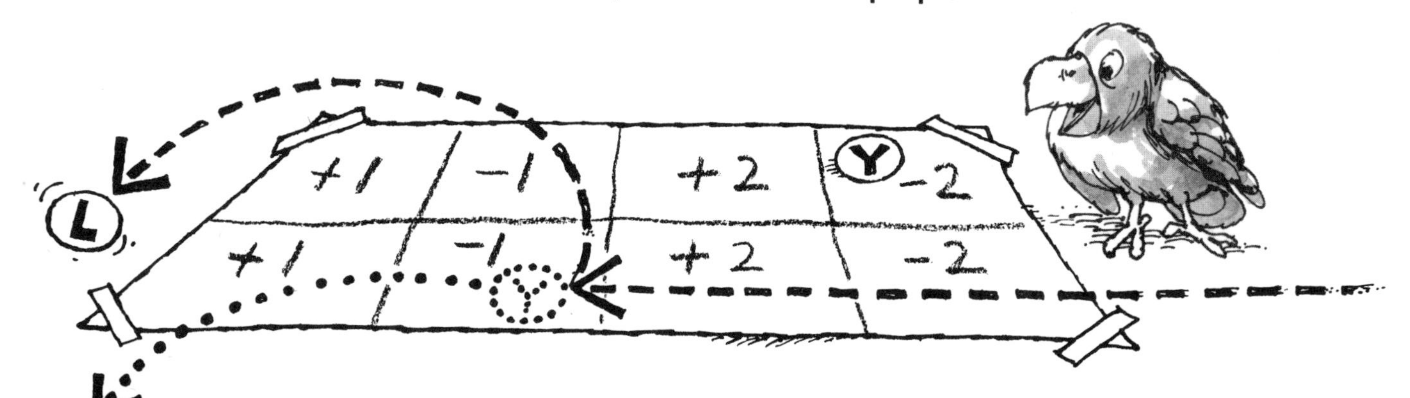

What is your score now? You can easily see that it is -2, because your other counter still is in the -2 square. But do you see what happened when Laura knocked your first counter out of the -1 square?

That subtracted -1 from your score of -3. When you subtract -1 from -3, you get -2. It is as if you add $+1$. Or it is as if you pay part of your debt.

This is the way it always works when you subtract a negative number from a negative number. What you are subtracting from is a debt.

There are many other ways to play the game of P.A.M. You can change it to suit yourself. You can change the numbers in the squares to any numbers you like, so long as you keep half of them positive and half of them negative. Like this:

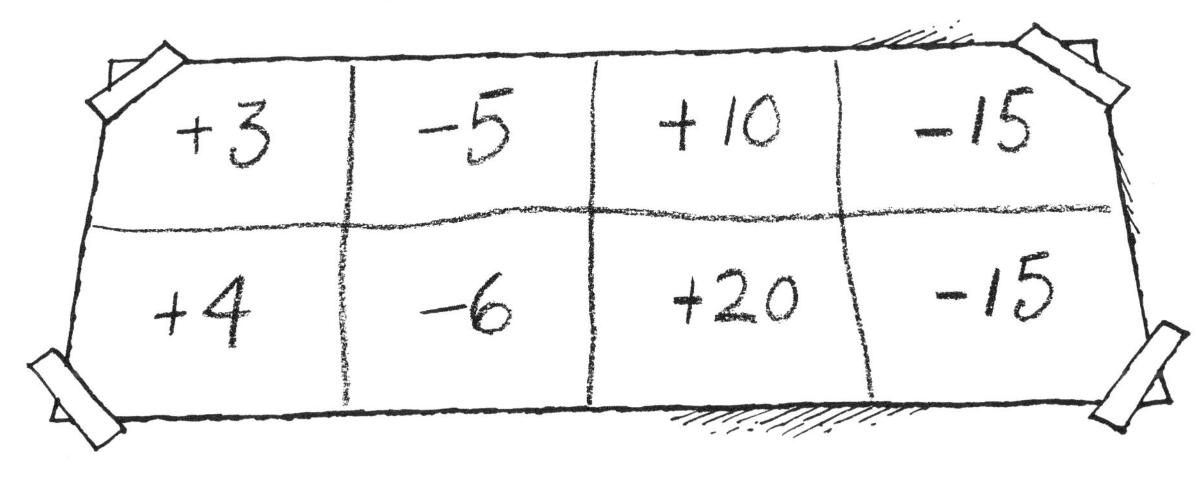

Or this:

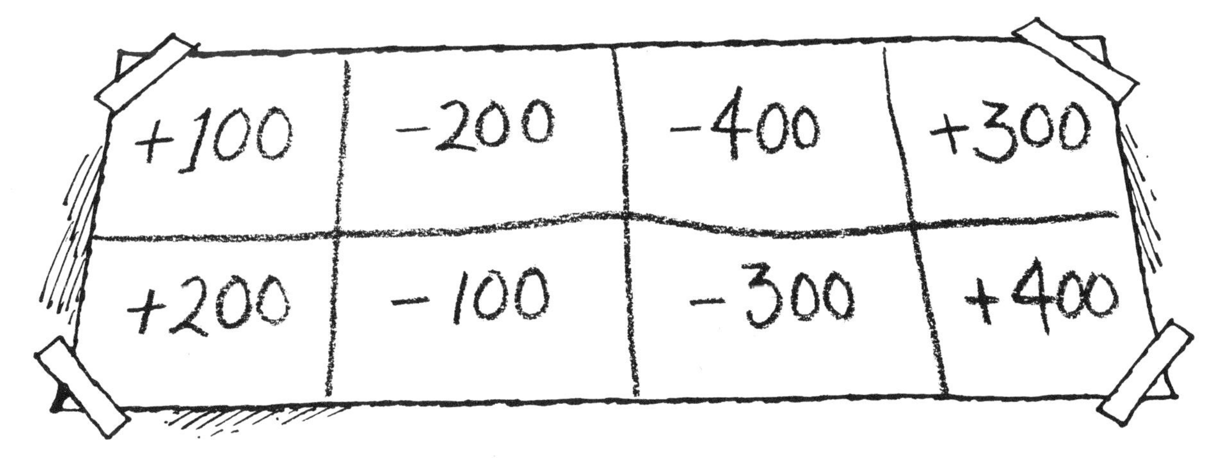

Or if you get tired of squares, you can draw circles or triangles or any other shapes you like:

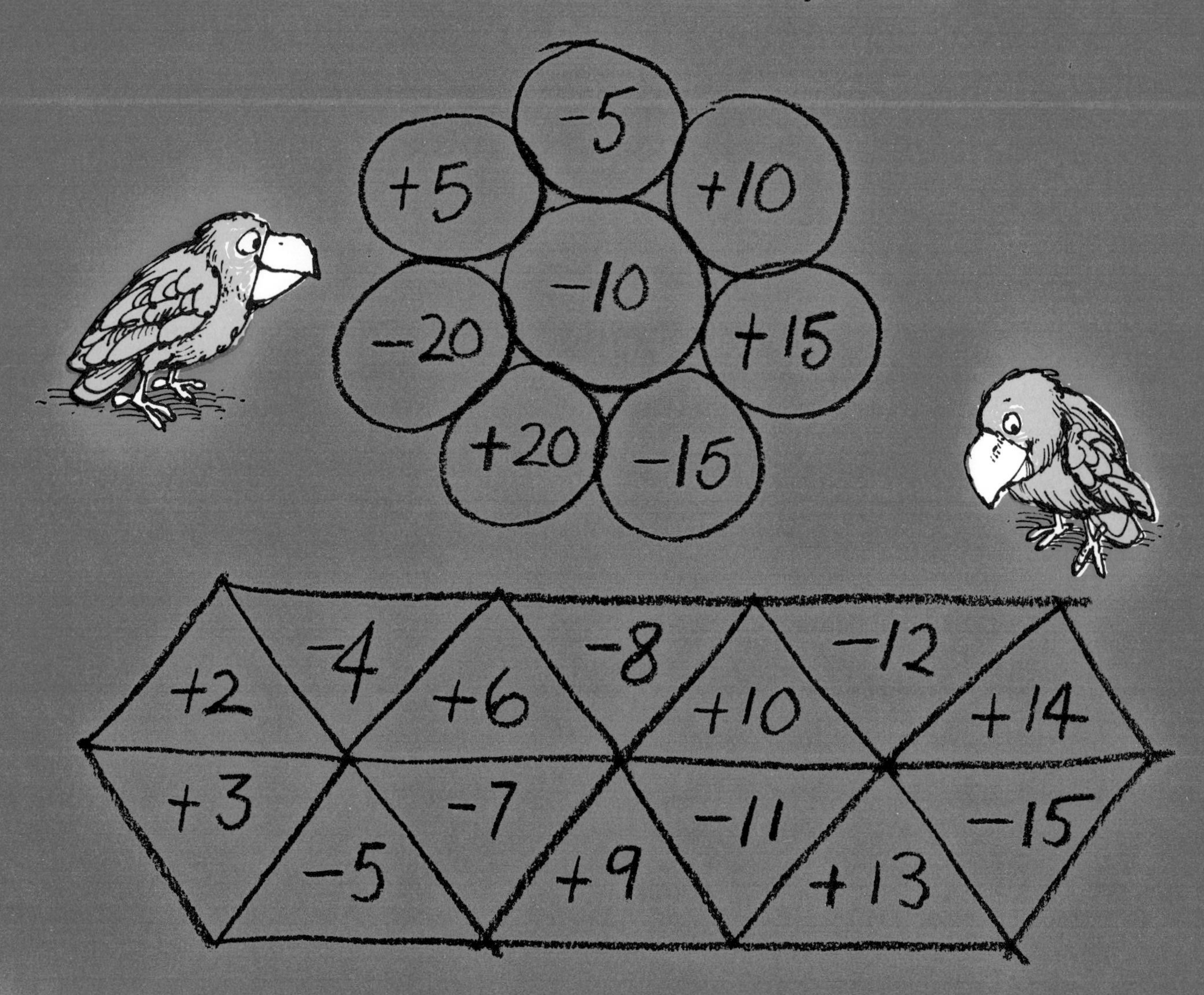

You can play the game outdoors, too. Draw whatever shapes you like in the dirt and mark the numbers on them. When you play it this way, it is best to have something round that you can roll—marbles or rubber balls or baseballs—for counters. Stand a few feet away from the shapes you have drawn and roll the counters toward them. And instead of just two counters apiece, you can try playing with three or four apiece.

Can you tell whether L or Y has won the game that ends with the counters like this:

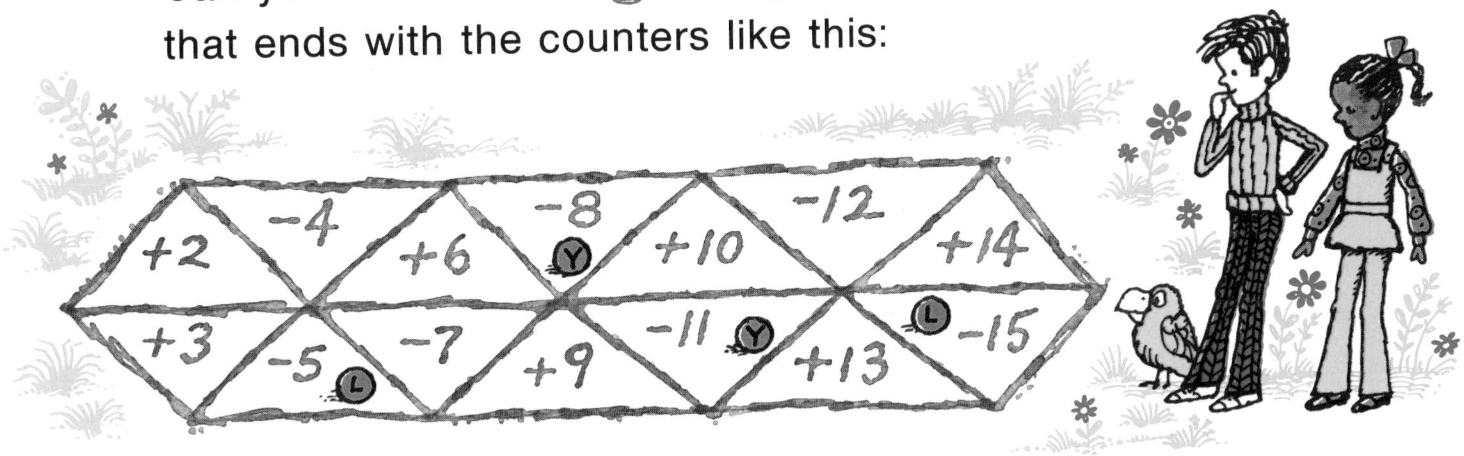

Have you guessed what P.A.M. stands for? It stands for Plus And Minus.

ABOUT THE AUTHOR

"If I hadn't made up my mind to be a writer, I would have been a mathematician," Robert Froman tells us. In this book Mr. Froman has combined his two great interests.

A free-lance writer since 1945, Mr. Froman first wrote mostly for magazines, and has been writing books for young readers since 1960. He has now published more than twenty books, most of them for children.

Robert Froman and his wife, Elizabeth Hull Froman, who is also an author of children's books, live in Tomkins Cove, New York.

ABOUT THE ILLUSTRATOR

Don Madden attended the Philadelphia Museum College of Art on a full scholarship. Following graduation, he became a member of the faculty as an instructor in experimental drawing and design. The recipient of gold and silver medals at the Philadelphia Art Directors' Club exhibitions, Mr. Madden's work has been selected for reproduction in the *New York Art Directors' Annual,* in the international advertising art publication, *Graphis,* and in the *Society of Illustrators Annual.*

Don Madden has always loved animals and the outdoors. He is delighted, therefore, to be living in an old house in upstate New York where he shares country life with his wife, an artist also, his two children, a golden retriever, 25 assorted chickens and roosters, a Nubian goat, and a black sheep named Josephine.